TRANSPORTATION EXPLORER

Can Military Trucks Float?

QUESTIONS AND ANSWERS ABOUT MILITARY VEHICLES

by Heather E. Schwartz

PEBBLE
a capstone imprint

Published by Pebble, an imprint of Capstone
1710 Roe Crest Drive, North Mankato, Minnesota 56003
capstonepub.com

Copyright © 2025 by Capstone. All rights reserved. No part of this publication may be reproduced in whole or in part, or stored in a retrieval system, or transmitted in any form or by any means, electronic, mechanical, photocopying, recording, or otherwise, without written permission of the publisher.

Library of Congress Cataloging-in-Publication Data
Names: Schwartz, Heather E., author.
Title: Can military trucks float? : questions and answers about military vehicles / by Heather E. Schwartz.
Description: North Mankato, Minnesota : Capstone Press, an imprint of Capstone, [2025] | Series: Transportation explorer | Audience: Ages 5–8 | Audience: Grades K–1 | Summary: "Tanks, helicopters, and fighter jets—the military has lots of amazing vehicles! How do they work, and what kinds of jobs can they do? Kids will find the answers to all their questions about military vehicles in this fun, interactive book"—Provided by publisher.
Identifiers: LCCN 2023043906 (print) | LCCN 2023043907 (ebook) | ISBN 9780756583125 (hardcover) | ISBN 9780756583071 (paperback) | ISBN 9780756583088 (pdf) | ISBN 9780756583095 (epub) | ISBN 9780756583101 (kindle edition)
Subjects: LCSH: Vehicles, Military—Juvenile literature. | CYAC: Military vehicles.
Classification: LCC UG615 .S45 2025 (print) | LCC UG615 (ebook) | DDC 623.74—dc23/eng/20231117
LC record available at https://lccn.loc.gov/2023043906
LC ebook record available at https://lccn.loc.gov/2023043907

Editorial Credits
Editor: Christopher Harbo; Designer: Terri Poburka; Media Researcher: Svetlana Zhurkin; Production Specialist: Katy LaVigne

Image Credits
Allied Joined Force Command Naples: Finnish Defense Forces Combat Camera/Ville Multanen, 32 (top); Getty Images: Archive Photos/Graphic House, 31 (top), Blade_kostas, 21 (tank), DarthArt, 13 (truck), JohnGomezPix, 20 (bottom), NickyBlade, 9 (truck), Placebo365, 29 (back), Rockfinder, 3 (bottom left), RoyalFive, 19 (tank), rusm, 4 (top), studiogstock, 21 (help sign), the8monkey, 21 (muddy puddle); NASA: Jim Ross, 31 (bottom); Shutterstock: Aditya0635, 17 (fighter jets), Africa Studio, 11 (bottom right), Andrey_Kuzmin, 19 (balloons), art4all, 4 (back), Ary Prasetyo, 5 (tank), Beautiful landscape, 27 (ship), BoeingMan777, 11 (back), 13 (smoke trail), Elena Dijour, 15 (back), fotoslaz, 7 (gas station), irin-k, 7 (back), Kanate, 23 (back), Kateryna_Moroz, 9 (donut in the pool), Luis Molinero, 4 (bottom), makarenko7, 29 (tank), Maria Martyshova (background), back cover and throughout, Marynova 13 (airplane wings), Maximillian cabinet, 12, oleskalashnik, 7 (drone), Pedal to the Stock, 29 (hard hat), rsooll, 19 (back), Sanit Fuangnakhon, 25 (helicopter), sdf_qwe, cover (floatie), Serg64, 13 (back), Sven Hansche, 23 (middle), Tavarius, 25 (back), Tracey Jones Photography, 27 (back), Triff, cover (sky), VanderWolf Images, 15 (fighter jet), 25 (fighter jet), VladisChern, cover (water); U.S. Air Force: Senior Airman BreeAnn Sachs, 18 (bottom), Senior Airman Haley Stevens, 18 (top), Senior Airman Matthew Seefeldt, 8, Senior Airman Ryan Callaghan, 17, Staff Sgt. James Merriman, 32 (middle), Staff Sgt. John Bainter, 16, Staff Sgt. Thomas Trower, 3 (bottom right), Tech. Sgt. Marleah Cabano, 26; U.S. Army: 30 (top), Capt. Scott Walters, cover (bottom right), Sgt. 1st Class Chad Menegay, 3 (top), Sgt. Sarah D. Williams, cover (truck), Spc. Bobby Allen, 5, Spc. Dustin Biven, 6 (bottom), Staff Sgt. Daryl Bradford, 6 (top), Staff Sgt. Jason Ragucci, 30 (bottom); U.S. Marine Corps: Cpl. Alexander Mitchell, 20 (top), 22, Cpl. Austin Gillam, 10 (top), Cpl. Cameron Hermanet, 3 (middle left), Lance Cpl. Betzabeth Y. Galvan, 14, Lance Cpl. Tyler Andersen, 3 (middle right), Sgt. Courtney White, 10 (bottom); U.S. Navy: Chief Mass Communication Specialist Darryl Wood, 31 (middle), James Mitchell, cover (bottom left), Lt. Courtney Callaghan, 28, Mass Communication Specialist 1st Class Ryre Arciaga, 32 (bottom), Mass Communication Specialist 3rd Class Aaron T. Smith, cover (bottom middle), Petty Officer 3rd Class Riley McDowell, 24

Any additional websites and resources referenced in this book are not maintained, authorized, or sponsored by Capstone. All product and company names are trademarks™ or registered® trademarks of their respective holders.

Printed and bound in China. 5834

Military vehicles do a lot of work. They carry cargo, equipment, and people too. They move along the ground, fly through the air, and float on water.

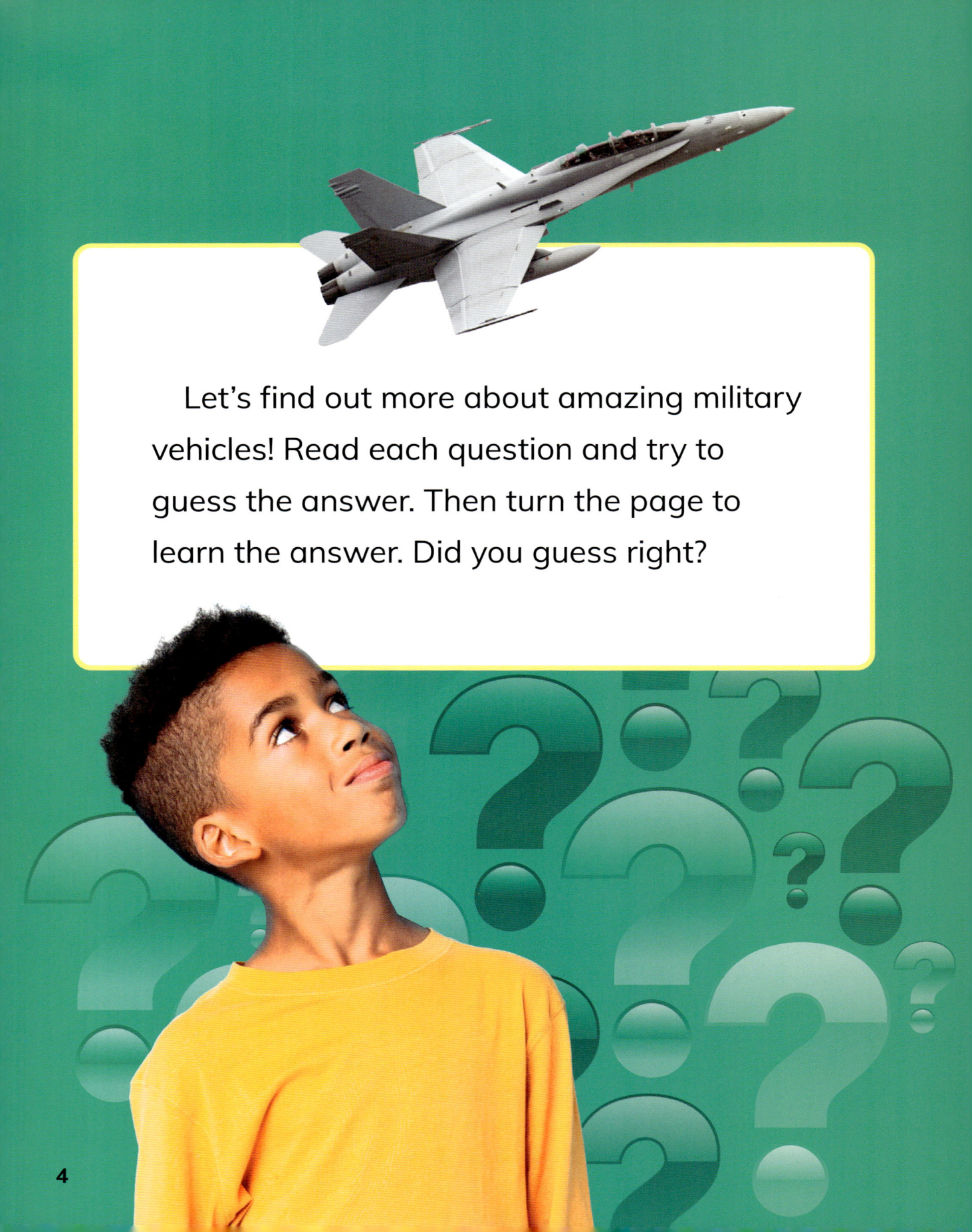

Let's find out more about amazing military vehicles! Read each question and try to guess the answer. Then turn the page to learn the answer. Did you guess right?

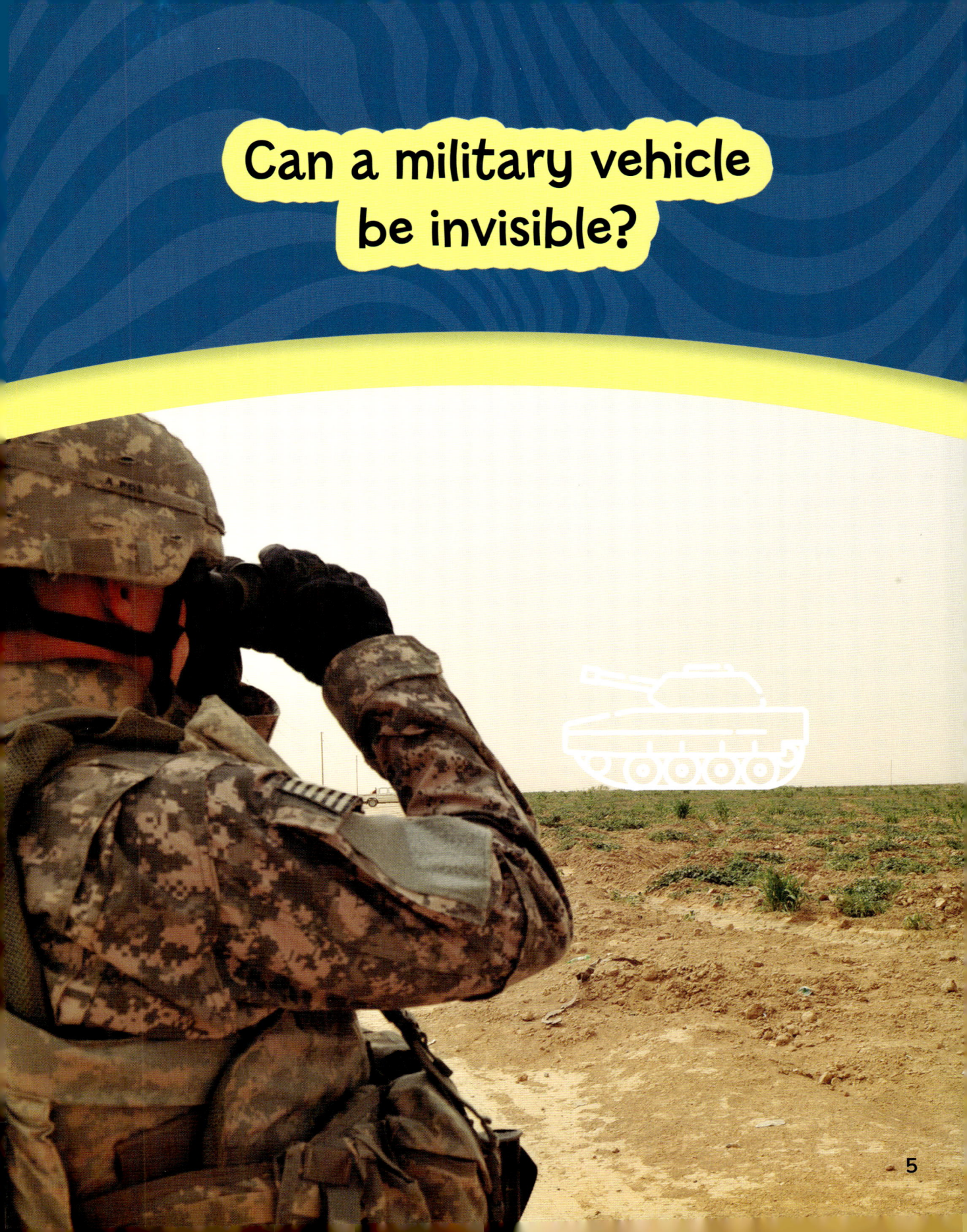

Can a military vehicle be invisible?

Not exactly. But military vehicles can hide very well! Some are painted light brown and tan to blend into deserts. Others are green and brown to look like forests. Their color patterns make military vehicles hard to see in their surroundings.

Can a military gas station fly?

Gas stations don't fly. But military refueling planes do. They fill the fuel tanks of other planes in midair!

refueling plane

How does it work? While flying, the refueling plane connects a long hose to the other plane. Fuel is then pumped from one plane to the other. When the tank is full, the planes disconnect and fly away!

Can military trucks float?

The Amphibious Combat Vehicle (ACV) sure can! This military vehicle has wheels to drive on land. But it can also go straight into water and float. Then two propellers move it forward like a boat. Soldiers use ACVs on lakes, rivers, and even oceans.

What is the fastest fighter jet in the world?

The MiG-25 Foxbat wins the race! It can fly 2,190 miles (3,524 kilometers) per hour. That's more than twice the speed of sound! When it flies overhead at that speed, it creates a thundering sonic boom.

Can military trucks fly?

Military trucks can fly with the help of heavy lift helicopters. These massive choppers move trucks, tanks, and even airplanes from one place to another.

Heavy lift helicopters can carry 22 tons (20 metric tons) of cargo. That's like lifting about three African elephants!

Can a military plane take off or land without a runway?

Some military planes don't need a runway. They take off and land by flying straight up and down!

The CV-22 Osprey takes off by flying straight up like a helicopter. Then its rotors tilt down so it can fly like an airplane. To land, the Osprey's rotors tilt back up. The aircraft flies straight down to the ground.

Can military planes fly all by themselves?

Military drones fly without human pilots onboard. Drones use computers and robotics to move through the air. Sometimes, pilots on the ground fly drones with remote controls.

MQ-9 Reaper drone

drone control center

Drones use cameras and other equipment to watch areas below them. They can fly places that are too dangerous for people to go.

What keeps heavy tanks from sinking in the mud?

Tanks use tracks instead of tires to move through mud. The tracks spread a tank's heavy weight evenly. They help keep the tank from sinking.

Tracks also have more gripping power. They give tanks better traction to keep moving over mucky ground.

What happens if a tank gets stuck?

21

Military recovery vehicles rescue tanks in trouble! These vehicles are so tough they can move an 80-ton (73-metric-ton) tank.

Military recovery vehicles can hoist tanks off rocks and pull them out of mud. They can even turn flipped tanks right side up!

Are there airports in the ocean?

Yes! Aircraft carriers are ships that serve as floating airports. Fighter jets and helicopters use them to take off and land at sea.

Gerald R. Ford class aircraft carriers are the largest in the world. These massive ships can hold more than 75 military aircraft!

Can a military plane carry a helicopter?

The C-5 Galaxy can fit up to five military helicopters in its cargo area. It is the largest aircraft in the U.S. Air Force.

How does cargo get in and out of this giant plane? The C-5's nose flips up to open and tips down to close. It looks like the plane has a giant mouth!

Do military ships dive underwater?

Military submarines do! These ships can dive deep below the ocean's waves.

What makes subs sink or float? A sub sinks when water is pumped into special tanks. To rise back to the surface, air is pumped into the same tanks to push the water out. The air makes the sub float.

Can military vehicles build bridges?

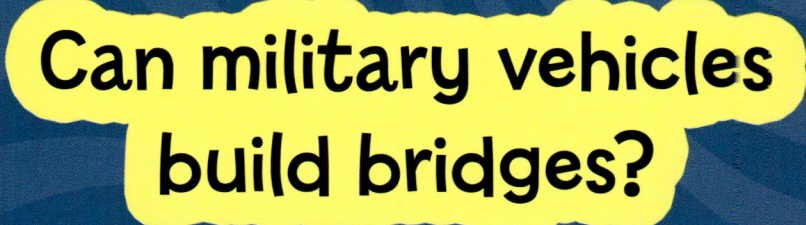

29

The M104 Wolverine can lay down a bridge! This armored bridge layer carries a bridge that can unfold in less than five minutes! Soldiers use the M104 whenever an instant bridge is needed. Its bridge can cross canals, trenches, and muddy ground.

FUN MILITARY VEHICLE FACTS!

The Wright Military Flyer of 1909 was the first military airplane. It held a pilot and one passenger.

Life on board a military submarine is so crowded there aren't enough beds for everyone. Sailors must take turns sleeping in the bunks.

The SR-71 Blackbird can fly 85,000 feet (25,908 meters) above the ground. That's more than twice as high as regular passenger planes can fly.

Camouflage comes in many different colors. Gray, black, and white are good for blending into city streets and snowy landscapes.

Military planes are often painted gray. The color helps them blend in with hazy, gray skies.

Our planet has more than 40,000 airports. But only 46 aircraft carriers sail Earth's seas. The United States owns 20 of them.